Understanding the Elements of the Periodic Table™

TIN

Janey Levy

rosen publishing's rosen central

New York

Published in 2009 by The Rosen Publishing Group, Inc.
29 East 21st Street, New York, NY 10010

First Edition

Library of Congress Cataloging-in-Publication Data

Levy, Janey.
Tin / Janey Levy.
p. cm.—(Understanding the elements of the periodic table)
Includes bibliographical references and index.
ISBN-13: 978-1-4358-5073-6 (library binding)
1. Tin. 2. Periodic law—Tables. I. Title.
QD181.S7L48 2009
546'.686—dc22

2008019949

Manufactured in the United States of America

On the cover: Tin's square on the periodic table of elements. Inset: The atomic structure of tin.

Contents

Introduction

The term "chemical element" may first make you think about your science classroom and laboratory experiments. The elements certainly play a role there. However, their importance extends far beyond classrooms and laboratories. Did you know the elements have actually affected history?

Chemical elements have played a determining role in human events many times over the centuries. Metals have been especially important. You probably know how the 1848 discovery of gold (Au) in California caused thousands of people from around the world to rush there in hopes of getting rich. But it's not only precious metals such as gold that have affected history. One of the reasons the ancient Romans invaded Britain in 43 CE was for the tin found there. The Romans used tin to make vital alloys, including pewter and bronze. They used pewter for items such as official seals and eating utensils. Bronze was even more valuable. It was used for many items, including decorative ornaments, kitchen and eating utensils, and—most important—weapons.

Tin continues to be important in modern society and is used in a wide variety of ways. It's still used in bronze and pewter, as well as a number of other alloys. It's used in agents for killing crop pests, certain kinds of paint, car parts, food containers, dental fillings, and a whole host of everyday

Tin is a silvery gray, shiny metal. It rarely occurs that way in nature, however. It's usually found in ores, the most common of which is cassiterite, or tin oxide (SnO_2).

items. It's also used in the electronic devices that are so essential to life today, including computers, cell phones, microwave ovens, and even video games and portable music players! It would be almost impossible for you to go through an entire day without encountering many things that contain tin. Life as we know it would be very different without this metallic element.

Chapter One
Tin: One of the Oldest Known Metals

Tin's chemical symbol is Sn. If the name and symbol don't seem to match, it's because they come from different languages. "Tin" was the name used by the Anglo-Saxons, the people who took over Britain in the 400s and 500s CE. The symbol comes from the Latin name for the metal, *stannum*, which was used by the Romans.

Tin is one of the oldest metals known to humans. It's been used for at least 5,500 years. In ancient times, it was employed primarily in three alloys. Bronze, made of tin and copper (Cu), was used to fashion jewelry, art, tools, utensils, and weapons because it is much harder than tin or copper alone. Solder, made of tin and lead (Pb), has a low melting point and is

The Romans sometimes used tin for ornamental purposes, even on practical objects such as battle helmets. Silvery tin and golden bronze once decorated this first-century CE helmet.

used to join metal surfaces. It was employed in metal pots, tools, and Roman plumbing. Pewter, made of tin and small amounts of other elements, was used for Roman government seals and some eating utensils.

Today, tin is still used to make bronze, solder, and pewter, as well as newer alloys. The alloys and other forms of tin are employed in the manufacture of a range of items. An incomplete list includes tinplate, gunmetal, bearings, organ pipes, bells, plastics, marine paint, and biocides. Tin is even used in a special glassmaking process!

The History of Tin

We don't know who discovered tin, or where and when it was first used. However, surviving evidence indicates it was widely used in alloys in the ancient world. The earliest written records of the Mediterranean region mention tin. Archaeologists have discovered thousands of ancient bronze objects.

Bronze was so important in ancient times that a whole period of history is named for it. Although the dates differed among regions, the Bronze Age lasted from about 3500 BCE to about 1100 BCE. Weapons, armor, pots, pans, plates, drinking vessels, bells, jewelry, art, decorative articles, religious items, and other bronze objects have been found in Iraq, Turkey, Greece, Egypt, India, and China. The Phoenicians of ancient Lebanon are known to have crafted tin containers for use in the production of their renowned purple dye (called Tyrian purple). When Mediterranean tin supplies (primarily from the Taurus Mountains in Turkey) ran low, the Phoenicians searched for other sources. They discovered the "tin islands," the location of which they kept secret. Today, it is believed the "tin islands" were northwestern Spain and Cornwall (part of the British Isles).

Other tin alloys were used in ancient times as well. The Romans used solder to join the lead pipes in their plumbing systems. They also fashioned objects of pewter, as did the Greeks, Egyptians, and Chinese.

This picture from a book published in the Netherlands in 1849 gives us a glimpse into a nineteenth-century tinsmith or pewterer's workshop. Pictures of objects produced in the workshop decorate the border.

Europeans used pewter for kitchen and household utensils until the mid-1800s. They also employed tin to plate other metals that were used to make household utensils, such as kettles and basins. In addition, they crafted pots, pans, plates, cups, buttons, coins, and organ pipes from tin. Tin was even hammered into thin sheets to make tin foil, which was used for centuries before aluminum (Al) foil replaced it in the early 1900s.

From the mid-1800s to the early 1900s, tinplate was used to make decorative "tin" ceilings. Tinplate is a sheet of iron (Fe) or steel coated with a thin layer of tin. The iron or steel is stronger and cheaper than tin, but the tin resists corrosion (rusting) better than iron or steel. It was stamped with complex patterns based on the carved and molded plaster ceilings found in lavish European residences.

The 1800s were also a time when scientists learned more about this prized element. Around 1830, Swedish chemist Jöns Jakob Berzelius (1779–1848) conducted the first careful attempt to determine tin's atomic weight. He found it to be 117.64 (expressed as atomic mass units, or amu). French chemist Jean-Baptiste Dumas (1800–1884) more accurately measured tin's atomic weight a few years later, when he determined it to be 118.06. Dumas's measurement was very close to tin's currently accepted average atomic weight of 118.710.

Dumas was also one of many scientists in the 1800s who noted repeating patterns of properties among groups of elements. This observation ultimately led to the development of the periodic table.

Creating the Periodic Table

The periodic table is an organization system for the elements. The usefulness of such a system may seem obvious today, but this wasn't so clear when only a few elements were known. However, as scientists identified more and more elements, they acknowledged that such a system might exist. German chemist Johan Döbereiner (1780–1849) placed elements into triads, or groups of three, based on similar properties in 1817. About forty years later, Dumas extended this idea to suggest the existence of families with three to five elements in each. French geologist Alexandre-Émile Beguyer de Chancourtois (1820–1886) was the first to arrange all of the known elements in order of increasing atomic weight and by properties in 1862. This was the first true periodic table, although it looked quite different from the modern one.

Around 1865, British chemist John Newlands (1837–1898) realized that de Chancourtois's method of arranging the elements revealed a

John Newlands, shown here late in life, never felt he received the credit he deserved for his recognition of a pattern among the elements.

pattern: every eighth element had similar properties. Newlands believed he had discovered a basic law of chemistry. He called it the law of octaves because the interval in the pattern reminded him of octaves in music. As scientists identified more elements, it became clear that the specific interval Newlands noted didn't always apply. Yet, his recognition of a pattern with regular intervals was important. It was a fundamental feature of what was later called the periodic law.

According to the periodic law, arranging the elements in order of increasing atomic weight causes elements with similar properties to appear periodically, or at regular intervals. Russian chemist Dmitry Mendeleyev (1834–1907) and German chemist Julius Lothar Meyer (1830–1895) independently discovered the law in the late 1860s. In 1869, Mendeleyev used it to create a table of the known elements. His revised table (1871) had horizontal rows called *Reihen* (which means "rows," now called periods) and vertical columns called *Gruppe* (which means "groups"). Starting with the top row, Mendeleyev arranged the elements from left to right in order of increasing atomic weight. He began a new row each time he came to an element with properties similar to the first element in the row. As a consequence, the elements in each group, or column, had similar properties. Mendeleyev's table had gaps because the known elements of the time didn't completely fill it. He correctly believed scientists would discover elements that filled the gaps. The periodic law even enabled him to accurately predict the properties of a few of these still-undiscovered elements.

The modern periodic table is based on Mendeleyev's table, although it's ordered by atomic numbers (the number of protons in the nucleus of an atom of the element) instead of atomic weights. Even though Mendeleyev's table contained far fewer elements than the modern table, tin and most of the other elements in its group were in it. The group includes carbon (C), silicon (Si), germanium (Ge), tin, and lead. Like tin, carbon and lead have been known since ancient times. Berzelius had discovered silicon in 1824. Only germanium wasn't listed in Mendeleyev's 1869 table because it hadn't been discovered yet. However, its discovery in 1886 didn't surprise Mendeleyev. It's one of the elements whose existence he had predicted.

Chapter Two
Tin, Atomic Structure, and the Periodic Table

Atoms make up essentially everything with mass—including you! More than one hundred kinds of atoms exist. Each element is composed of only one kind. So, just what are atoms? They're minute bits of matter so tiny that you can't see them. To get an idea of how tiny they are, consider this: about 178,000 tin atoms side by side equal the diameter of a dot the size of a pin point (or 0.05 mm)! Yet, there are still smaller bits of matter called subatomic particles.

An Atom's Structure and Parts

Atoms aren't solid. They contain lots of space. A dense nucleus occupies the center and houses most of the atom's mass. It's composed of two kinds of particles: neutrons, which have no electrical charge, and positively charged protons. Together, they give the nucleus a positive charge.

Lightweight electrons orbit the nucleus. They're negatively charged, and when the number of electrons equals the number of protons, the atom as a whole has no electrical charge.

The orbiting electrons are organized in overlapping shells, or energy levels. Each shell can hold only a certain number of electrons. Shells farther from the nucleus can hold more electrons. The first shell can hold two, the second can hold eight, the third can hold eighteen, and the fourth can

Tin has two electrons in its first shell, eight in the second shell, eighteen in the third, eighteen in the fourth, and four in the fifth.

hold thirty-two. In theory, each remaining shell can hold many more electrons. But for all the known elements, these shells are never filled to capacity. All tin atoms contain fifty protons. Most have sixty-eight or seventy neutrons. Atoms with the same number of protons but different numbers of neutrons are called isotopes. In all, tin has ten stable isotopes. In neutral tin atoms, fifty electrons orbit the nucleus in five shells.

What If You Change an Atom's Structure?

What happens if you change the number of subatomic particles in an atom? Different changes would have different effects. Changing the number of electrons gives the atom an electrical charge and creates an ion of the original element. Adding an electron creates a negatively charged ion. Subtracting one creates a positively charged ion.

Changing the number of protons changes everything. The number of protons determines what element an atom is, so a different element would be formed if the number of protons were changed. If you could add a proton to the nucleus of a tin atom, you would have antimony (Sb). Like tin, antimony is a metal that's been known since ancient times. However, antimony is toxic and must be handled carefully. If you could subtract a proton from the nucleus of a tin atom, you would have indium (In), a rare, soft metal. Like antimony, it's toxic and requires careful handling. Because the number of protons is what determines an atom's identity, it has a special name. It's called

Tin  Snapshot

Chemical Symbol:	Sn
Classification:	Metal
Properties:	Soft, silvery-white, malleable, ductile, nonreactive with oxygen unless heated
Discovered by:	Known since ancient times
Atomic Number:	50
Atomic Weight:	118.710 atomic mass units (amu)
Protons:	50
Electrons:	50
Neutrons:	70, 68, 66, 69, 67, 74, 72, 62, 64, 65 (in stable isotopes, in order of decreasing abundance)
State of Matter at 68° Fahrenheit (20° Celsius):	Solid
Melting Point:	450°F (232°C)
Boiling Point:	4,716°F (2,602°C)
Commonly Found:	In Earth's crust

the atomic number. Thus tin, which has fifty protons, has an atomic number of 50. On the periodic table in this book (on pages 38–39), each element's atomic number appears to the upper left of the element's symbol.

Changing the number of neutrons alters the isotope of the element. Elements usually exist as several isotopes. Each isotope is generally identified by the element's name plus the isotope's mass number (the number of protons plus the number of neutrons). Tin has ten naturally occurring isotopes—more than any other element. The most common are tin-120 and tin-118. Tin-120 has fifty protons and seventy neutrons; tin-118 has fifty protons and sixty-eight neutrons. A number of man-made, radioactive isotopes of tin also exist.

Because each isotope has a different number of neutrons, each has a different atomic weight. The periodic table lists the average atomic weight of each element. An element's average atomic weight is determined by averaging the atomic weights of all its naturally occurring isotopes, taking into account the proportions in which they occur. Each element's average atomic weight appears to the upper right of the element's symbol on the periodic table in this book. Tin's atomic weight is 118.710 amu, which has been rounded to 119 in our periodic table.

What's That Sound?

Did you know that tin cries? The metal has a highly crystalline structure. Saying that an element has a crystalline structure means its molecules and atoms form a regularly repeating arrangement as a crystalline solid, often with a specific geometric shape. Tin's crystal structure is tetragonal, which basically means it's shaped like a 3-D rectangle. When a bar of tin is bent, the structure is disrupted, giving off a strange crackling sound known as the "tin cry."

What the Periodic Table Can Teach Us About Tin

You can use an element's location on the periodic table to learn about it. An element's place on the table tells you much about its properties.

One of the first things to look for is the staircase line on the right side. An element's place in relation to this line indicates whether it's a metal, nonmetal, or metalloid (an element that possesses some properties of both metals and nonmetals). To the left of the line are metals. To the right are nonmetals. Elements touching the line are metalloids except for aluminum and polonium (Po), which are metals. Tin is to the left, which tells you that it's a metal.

Several distinctive properties characterize metals. They're usually solids that can be polished until they shine. They conduct electricity and can be hammered into shapes (malleable) and stretched into wires (ductile).

Another obvious feature of the periodic table is its organization into rows and columns. The seven rows, or periods, are numbered from top to bottom along the left side. Each period's elements have the same number of electron shells, and the period's number indicates the number of shells. Tin is in period 5, so it has five shells that are completely or partially filled with electrons.

The eighteen columns, or groups, are numbered from left to right along the top. Two numbering systems are employed. An older system uses Roman numerals and letters of the alphabet, while the newer one uses Arabic numerals. As a general rule, all the elements in a group have the same number of electrons in their outermost unfilled shell(s) (with the exception of the noble gases in group 18, which have a full outer shell). That's significant because it's these electrons—called valence electrons—that do much to determine an element's chemical behavior. Because each group's elements (usually) have the same number of valence electrons, they behave similarly.

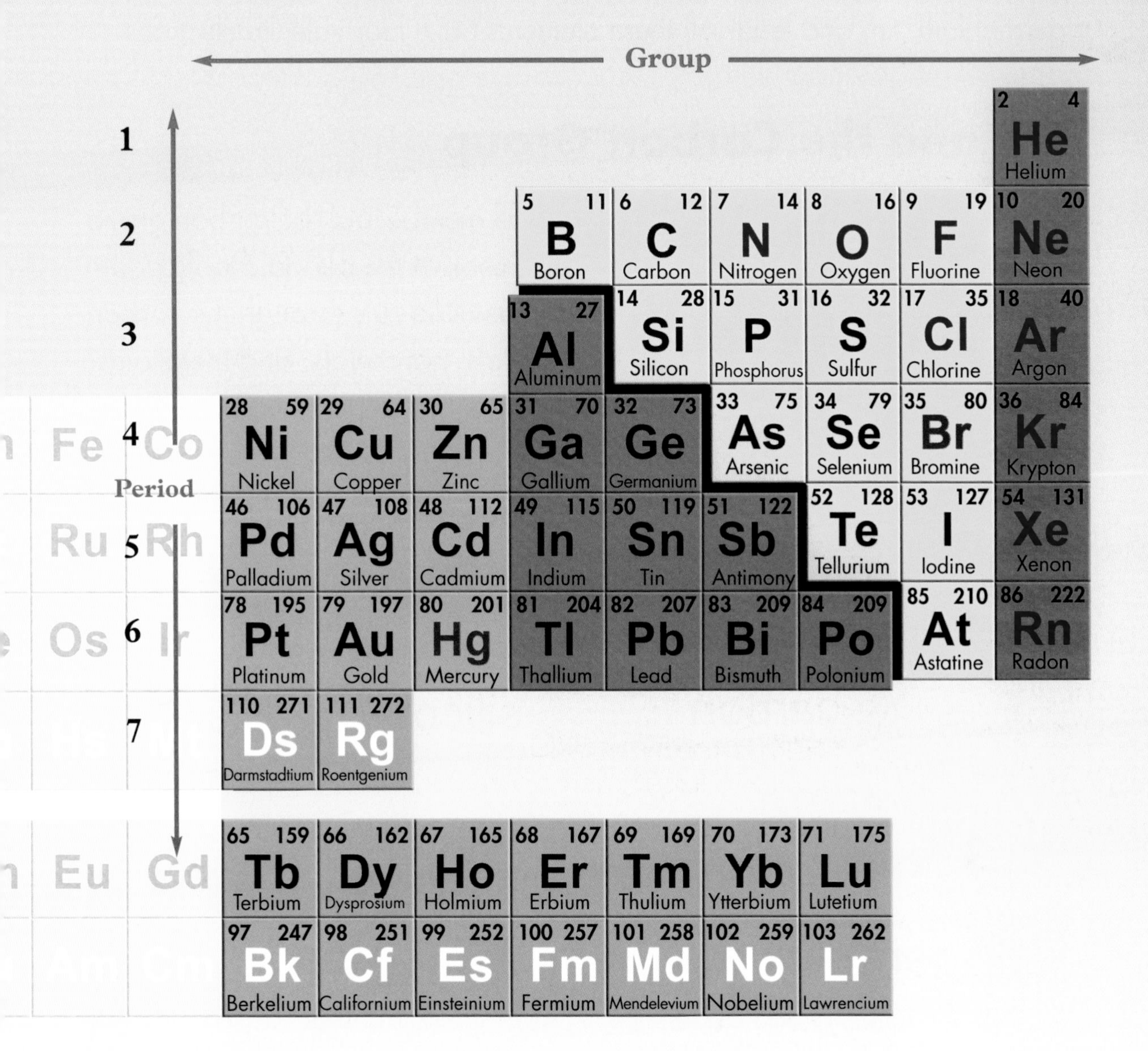

Period numbers run down the left side of the periodic table. Group numbers run across the top. You can quickly locate tin—or any element—using its period and group number. Tin's square is located in period 5 and group 14.

Tin is in the group numbered IVA in the traditional system and 14 in the newer one. The group is sometimes called the carbon group because carbon is the first element. In order from the top, the group contains carbon, silicon, germanium, tin, and lead. All these elements have four valence electrons.

Tin and the Carbon Group

Because tin is a metal, you might expect all members of the carbon group to be metals. However, that's not the case. Look at the periodic table again (see pages 38–39). Notice how the staircase line cuts through the carbon group. As a result, the group contains metals, nonmetals, and metalloids.

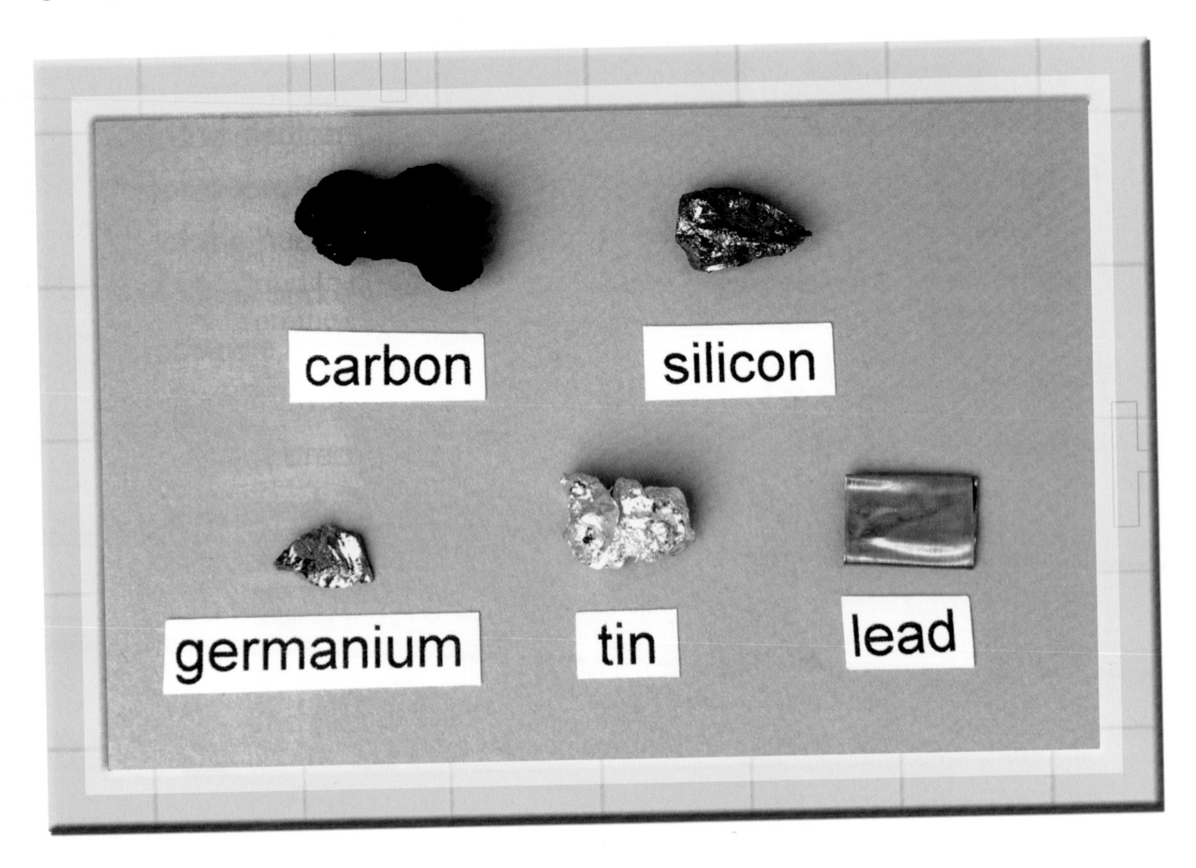

The elements of the carbon group, shown here, are all solids. The nonmetal carbon lacks the shininess of metal. The metalloids silicon and germanium are shiny, but they're also brittle.

Carbon is a nonmetal. Silicon and germanium are metalloids. Lead, like tin, is a metal. You may wonder why such varied elements are in the same group. But you already know the answer—they all have four valence electrons.

Having four valence electrons means the elements in the carbon group are reactive. That's because their outer shell isn't full. Except for hydrogen (H) and helium (He) atoms—which have a single shell that has one electron in hydrogen's case and two electrons in helium's case—atoms of many other elements are considered to have a full outer shell with eight electrons. An atom with a full outer shell is most stable, or unlikely to react further. Elements whose atoms have fewer than eight valence electrons usually undergo reactions to try to reach that number. That's why the members of the carbon group are reactive. However, they don't all react in the same way. Some give up electrons, some share electrons, and some gain electrons. But they're all reactive.

Chapter Three
Getting and Using Tin

Tin has two major solid allotropes, or crystalline forms. White, or beta, tin is the more familiar shiny metallic form. At temperatures below about 56° Fahrenheit (13° Celsius), beta tin slowly changes into a gray powder called gray or alpha tin, which has few uses. This transformation is called "tin disease," "tin pest," or "tin plague." Adding a small amount of antimony or bismuth (Bi) prevents the conversion.

Tin makes up only about 0.0002 percent of Earth's crust. Elemental tin—that is, tin alone—rarely occurs. Tin is mostly found in ores such as cassiterite (SnO_2). The main mining areas are in the "tin belt," which runs from China through Thailand, Malaysia, and Indonesia. Other important deposits occur in Bolivia, Brazil, Peru, Nigeria, and Democratic Republic of the Congo. However, supplies of tin ore won't be easily harvested forever. *New*

The term "cassiterite" comes from the name of the mysterious islands where the Phoenicians found tin. They were called the Cassiterides, from a Greek word meaning "tin islands."

The Devil's Work

Odd examples from history illustrate the problem of tin disease, the slow transformation of the stronger metallic beta tin form to the crumbly alpha tin at low temperatures. Centuries ago, northern Europeans noticed tin organ pipes crumbling in their unheated churches. They didn't understand what was happening and believed it was the devil's work. Other tin objects such as tin ingots, or bars, also crumbled during cold winters. However, by melting the gray powder and then re-cooling the melted tin to a solid, the ingots can be restored!

Scientist has estimated we'll have mined all the readily accessible tin by 2050, while environmentalist Lester Brown has calculated this will instead occur by 2030.

Obtaining Tin

Most of our tin comes from cassiterite. Smelting plants transform the tin in the ore into tin metal. Smelting uses temperatures between 2,192°F (1,200°C) and 2,372°F (1,300°C) to react the ore with carbon to produce molten tin metal. Some cassiterite is pure enough be smelted without any preliminary steps. However, some tin ores contain impurities that must be removed first. To accomplish this, the ore is roasted then leached with acid or water.

For smelting, a batch of ore, called a charge, is placed in a reverberatory furnace, which is heated. The charge may contain scrap or recycled tin in addition to the ore. Because the ore still contains some impurities,

With a power drill, a miner breaks up ore-bearing rock in the Geevor tin mine in Cornwall, England. Such underground mining is difficult, dangerous, and expensive. Competition from cheap tin from Asia forced Geevor to close in 1990.

substances are added to remove them. Carbon takes away oxygen. Limestone ($CaCO_3$) and silica (SiO_2) fluxes remove other impurities.

After smelting, the molten charge is poured into a settler. The denser tin settles to the bottom. The less dense slag—the combined fluxes and impurities—rises to the top and is poured off. The slag may be resmelted to recover any remaining tin. The molten tin is cast into slabs or pigs. It's refined and then cast into ingots, or bars, that weigh about 100 pounds (45 kilograms) each. Now, it's ready to be used.

Uses of Unalloyed Tin

Because tin is relatively soft, the unalloyed metal has few applications. However, one important use is in the manufacture of tinplate, or tin-plated metals. Worldwide, more than 16 percent of tin went into tinplate in 2006, according to ITRI, Ltd., a group that represents tin producers. In turn, more than 90 percent of tinplate is used for containers. This is because of two important properties of tin: it's nontoxic and resists corrosion. Tinplate cans (tin-plated steel)—often incorrectly called "tin" cans—are used for a range of items, including food, cleaning products, polishes, and spray products. The seams of these cans may also be soldered with unalloyed

You might be surprised how long "tin" cans have been used for food. The first food-canning factory that used "tin" cans opened in England in 1811.

tin. Tinplate containers are used for fuels as well. In addition, tinplate is used in signs, decorative "tin" ceilings, and food-preparation containers and utensils.

Tin's malleability allows manufacturers to make it into extremely thin foil. This foil blocks the passage of moisture and can be used as a wrapping material. However, because of the higher cost of tin, aluminum foil is now commonly used instead of tin foil.

Unalloyed tin also has a use that may surprise you—it's often employed to manufacture glass. In a special method called the Pilkington process, molten glass is poured onto a pool of molten tin and allowed to cool and harden. Because the glass floats on the pool of molten tin, it's called float glass. This method produces glass with very smooth, flat surfaces. Most window glass is made this way.

Uses of Tin Alloys

Manufacturers make a number of useful tin alloys today, including some that have been made for thousands of years. Solders are perhaps the most important. Worldwide, they accounted for about 52 percent of all tin used in 2006, according to ITRI, Ltd. Solders are used in all sorts of products: cans, plumbing, sheet metal work, cables, automobile radiators and body parts, and electronic and electrical products. Traditionally, solders were made of tin and lead. That's changing, though, as countries ban the use of lead in electronic devices because of its toxicity. New solder alloys without lead, such as one combining tin with silver (Ag) and copper, have been developed.

Besides solder, other alloys of tin and lead have several applications. They're used to make organ pipes, with the amount of tin varying according to the tone desired. A tin-lead alloy called Wood's metal is used in fire-sprinkler systems. This alloy is a solid at room temperature but melts at

Tin Whiskers

Tin whiskers are tiny, hair-like crystals that grow without warning from tin. They present a special problem in electronic devices, where tin finishes and solder are widely used. The whiskers can grow between electrical components, causing a short in the system. They've caused problems in heart pacemakers, radar systems, communications satellites, nuclear power plants, and the space shuttle.

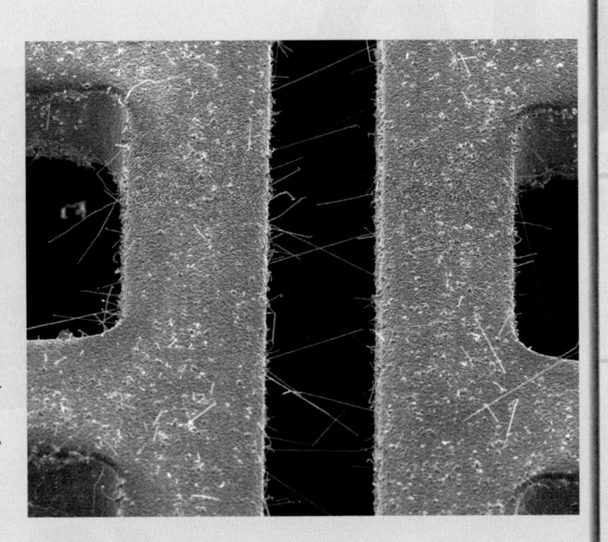

This greatly enlarged photograph shows a tin-plated copper part used in electronic devices. When it was removed from storage after three years, it was covered with fine tin whiskers.

about 160°F (70°C) to open a valve to release water. The alloy's low melting point makes it ideal for this use. The low melting point of tin-lead alloys makes them a good choice for safety plugs in steam boilers, too. If too much steam builds up inside a boiler, it will melt the plug, allowing steam to escape before the pressure gets so great that it bursts the boiler. A tin-lead alloy is also used to coat sheets of iron to produce terneplate, which is used to make roofing, fuel tanks, and fire extinguishers.

Like solder, the ancient alloys bronze and pewter are still produced. Other modern tin alloys include gunmetal (tin, copper, lead, and zinc [Zn]) and bell metal (tin and copper). Tin is sometimes added to brass (an

Tin-lead solder wire is available with varying tin-to-lead ratios. Solder with a high tin content melts at lower temperatures and is less likely to crack. It's also more expensive to buy.

alloy of copper and zinc) to make it more resistant to corrosion. It's combined with antimony, copper, and sometimes lead to make Babbitt metal, which is used to make bearings employed in trains. Tin is sometimes added to titanium (Ti) alloys to strengthen them. Tin may also be added to cast iron. Adding as little as 0.1 percent of tin makes cast iron stronger, more resistant to wear, and easier to work.

Tin alloys are used in die-making and die-casting. A die is a mold, and die-making is the production of molds. Tin alloys can be used to make molds employed in the low-volume production of plastic parts. Die-casting is the production of objects or machine parts by forcing molten metal into dies and allowing it to harden. Because tin alloys have a low melting point and flow easily, they're especially good for making items or parts that have lots of fine detail.

Tin and niobium (Nb) form a crystalline alloy that's superconductive at very low temperatures. This may help to create very strong superconductive magnets that use almost no power.

In addition to the uses listed here, tin forms several compounds that have a wide range of applications.

Chapter Four
Tin Compounds

Compounds are made up of two or more elements that are bonded to one another. Bonding is usually characterized as either covalent or ionic. In covalent compounds, atoms share valence electrons. In ionic compounds, positively charged ions (such as an atom of an element that has lost one or more electrons) and negatively charged ions (such as an atom of an element that has gained one or more electrons) are attracted to one another. Tin gives up electrons when it reacts with other elements, so it forms ionic bonds. Tin atoms don't always give up the same number of electrons when they form bonds. Sometimes, a tin atom gives up two electrons; other times, it gives up four. Because of this, scientists say that tin has two oxidation states. The term "oxidation state" is used to describe how many electrons an atom has lost (or gained). Scientists indicate an element's oxidized state by attaching a number equal to the number of electrons lost (or gained). So, a tin atom that's lost two electrons is tin(II); one that's lost four electrons is tin(IV). Tin forms both inorganic and organic compounds. Let's look at inorganic tin compounds first.

Inorganic Tin Compounds

An inorganic compound is any compound that doesn't contain carbon and hydrogen. Tin forms a number of inorganic compounds whose uses

Do you want to know if your favorite fruit juice contains tin(II) chloride? Read the label to find out. Remember, it might be listed as stannous chloride, rather than tin(II) chloride.

range from food additives to fire retardants.

Tin forms two compounds with chlorine (Cl). Tin(II) chloride ($SnCl_2$)—sometimes called stannous chloride—is a food additive used in canned and bottled foods and fruit juices. It keeps the food or juice fresh longer because its ready reaction with oxygen prevents the oxygen from reacting with the food itself and reducing its freshness. Tin(IV) chloride ($SnCl_4$), or stannic chloride, is used as a mordant in dyeing some fabrics. The mordant attaches to the fabric, and the dye attaches to the mordant. Moreover, tin(IV) adds substance to thin, lightweight fabrics such as silk.

Tin(IV) vapor is used to strengthen glass. When applied to freshly made glass, it leaves behind a very thin, transparent film of tin(IV) oxide (SnO_2) that makes it stronger. This allows manufacturers to make glass thinner without any loss of strength.

Tin(IV) oxide has several additional applications. It's used in sensors to detect certain dangerous gases. Its electrical conductivity increases as it absorbs the gases, and this can be monitored. Tin(IV) oxide also forms a whitish powder that's used to polish marble and decorative stones. The powder can be mixed with other compounds to form colorful glazes for pottery.

Tin also combines with numerous elements to form stannate compounds that have varied uses. For example, it combines with cobalt (Co) to produce cobalt stannate (Co_2SnO_4), a beautiful sky-blue pottery glaze. Tin and zinc form zinc stannate (Zn_2SnO_4), a fire retardant with the additional desirable property of reducing the amount of smoke that is produced.

Organotin Compounds

An organic compound is one that contains carbon and hydrogen. Tin forms numerous organic compounds, and they have at least as many or more uses as any other metal's organic compounds.

Organic tin, or organotin, compounds are mainly used as stabilizers for plastics made of polyvinyl chloride, or PVC. Heat, light, and oxygen break down PVC plastics, leaving them discolored and brittle. Adding organotin compounds inhibits this decay and makes possible the many items made of PVC plastics. These items include the covering on electric wires, water and sewer pipes, siding for houses, flooring, toys, food packaging, bottles, credit cards, and garden furniture.

Organotin compounds also play a role in the production of some silicones. Silicones are human-made materials that have an enormous number of applications. They can be vulcanized, or made rubbery, to create highly useful silicone rubber. One vulcanization process uses organotin

The many everyday PVC items that tin helps make possible may not be around much longer. Studies have linked PVC to numerous health problems, including cancer.

compounds. Silicone rubber made this way is employed for molds, adhesives, coatings for electronics components, and automobile gaskets (seals).

Another major application of organotin compounds is as agricultural biocides. These biocides protect crops primarily by killing insects and fungi. This application is an especially important use of organotin compounds because these pests destroy more than one-third of the world's food crops annually. These compounds are fairly selective in their action, and their decomposition products are widely regarded as nontoxic to mammals.

Special marine paints containing organotin compounds protect boats against barnacles and wood-eating worms. Similar products act as wood preservatives and prevent mold growth on stone structures.

Dangers of Organotin Compounds

Little research has been done on how organotin compounds may affect people. However, we do know some of them in sufficient quantities can make people sick, cause long-lasting damage to their bodies, and even kill them. We know more about their effects on animals because more research has been conducted in that area. Some studies have shown organotin compounds affect the immune and/or reproductive systems. Real-life events have supported the studies. Twenty years after marine paint containing an organotin compound was introduced, it was associated with the harming of some marine organisms. The compound appeared to be producing strange changes in oysters and other marine animals, causing some to become infertile. These discoveries caused countries to limit or ban the paint's use.

In water, organotin compounds have the potential to spread readily and harm entire ecosystems. They can be toxic to fish, fungi, algae, and phytoplankton (tiny plantlike organisms). Because algae and phytoplankton play important roles in aquatic ecosystems, harm to them has much larger consequences. Algae help purify the water, and phytoplankton provide

A sailor is painting the bow of an aircraft carrier. Because of the dangers posed by the organotin compound TBT, the U.S. Congress severely limited its use in marine paint in 1988, and the navy no longer uses paint containing TBT.

oxygen for other water organisms. Both are important food sources in aquatic food chains.

Once organotin compounds enter the environment, sunlight and bacteria break them down into inorganic tin compounds. However, for some compounds this can take weeks, allowing levels of the compounds to build up in fish, other organisms, and plants. If people eat these things, they're consuming the organotin compounds that they contain.

Scientists need to do more research to determine the effects of organotin compounds on people, animals, and the environment. Until we learn more, we must use these compounds with care and exercise caution in the ways we dispose of products made with them.

Chapter Five Tin and You

Tin isn't a common element, but it's important in modern life. We've already looked at some of the ways tin, tin alloys, and tin compounds are used. Today, you may have directly or indirectly encountered some of the products we discussed. Perhaps you ate food grown with the help of biocides containing organotin compounds. Or, maybe you ate food that came from a tin-plated steel can or from packaging made with organotin compounds. You may have had a drink from a bottle made of PVC plastic. Perhaps your school's or home's windows were made using the Pilkington process. Maybe your school has a fire-prevention sprinkler system made with a tin-lead alloy. However, these are only a few of the ways tin may play a role in your life.

Automobiles and Tin

Automobiles are the primary means of transportation for most people in the United States. You might even say automobiles are part of the American identity. And automobiles rely on tin.

Tin alloys are employed in many ways on both the outside and inside of automobiles. Terneplate—steel coated with a lead-tin alloy—is used for automobile frames, gasoline tanks, and some engine parts. Cast iron with tin added to increase its strength is used for some engine and brake parts

Robots work on cars on an assembly line. Tin may have a new application in "green" cars. It's used in safety sensors that detect hydrogen leaks in hydrogen fuel cell cars.

and for the crankshaft, which helps transfer power from the engine to the wheels. Tin alloys are employed as coatings in brake systems. Solders are used in automobile parts such as the radiator, which helps keep the engine from getting too hot. Large amounts of solder are employed to fill spaces along seams and welds in automobile bodies to ensure that the joints and surfaces are smooth. Even the manufacture of the windows involves tin. The Pilkington process is used to make glass for automobiles—and for school buses!

Electronics and Tins

Electronic devices dominate modern life. They're such a part of ordinary life that most of us use them without thinking much about them. The long list of familiar electronic devices includes computers, video games, cell phones, digital watches, DVD players, portable music players, portable e-mail devices, and microwave ovens. It also includes heart pacemakers and communications satellites. These last two devices may seem less connected with your life than the other items, but you might be surprised. Perhaps you have a grandparent who has a pacemaker that keeps his or her heart beating regularly. If you have satellite television or radio, you get signals

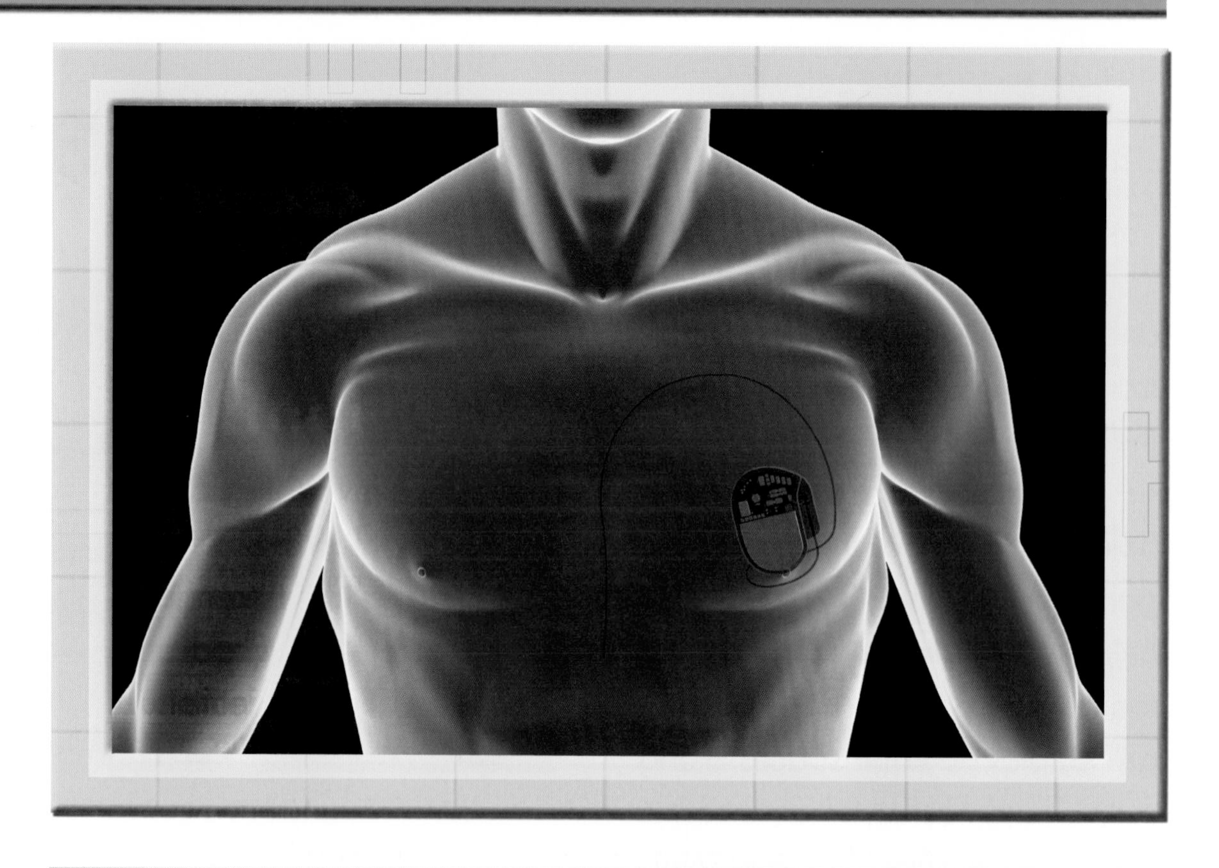

Pacemakers are important electronic devices that rely on tin. The doctor inserts the pacemaker—clearly visible in this X-ray—under the skin on the chest. It sends an electric signal to the heart through a wire.

from a communications satellite. The satellites also send signals for credit card approval when you use one of the pay-at-the-pump systems at a gas station. All the devices listed here depend on tin.

The major application of tin in electronic devices is in solders. One of the most important applications of solders is in the attachment of microchips to circuit boards, which form the heart of electronic devices. Microchips are tiny electronic circuits composed of even tinier parts. They carry out computer instructions or store information. Circuit boards are small boards with attached microchips and other electronic parts that are connected to form a complete circuit.

Tin in Your Mouth!

You may not know it, but you may have tin in your mouth. If you have fillings in your teeth, you have tin. The modern alloy for fillings contains silver, tin, and copper. In addition, tin is often employed in the gold alloy sometimes used to crown, or cap, teeth. It's most commonly used when the gold crown will have a porcelain, or ceramic, covering so it matches your other teeth. The tin helps the crown and covering stick together.

Solders aren't the only application of tin in electronic devices. Tin is used in coatings applied to the wires of the various parts on the circuit boards. In addition, it's employed in the coatings covering the pads on the boards to which the wires are soldered. So, next time you use your cell phone or play a video game, remember that tin makes it possible!

Tin at Home and in School

Even if you haven't ridden in an automobile or used an electronic device today, you've still probably encountered tin. Tin-plated steel cans are used for food, cleaning products, polishes, paint, and spray products. Tinplate containers are also employed for some medicines and makeup. In addition, tinplate is used to make batteries, toys, and baking equipment. And remember, solder is used on many of these items and on some plumbing as well.

Tin is applied to many ordinary metal items to give them a shiny coating. Did you use paper clips or staples at school today? They're usually

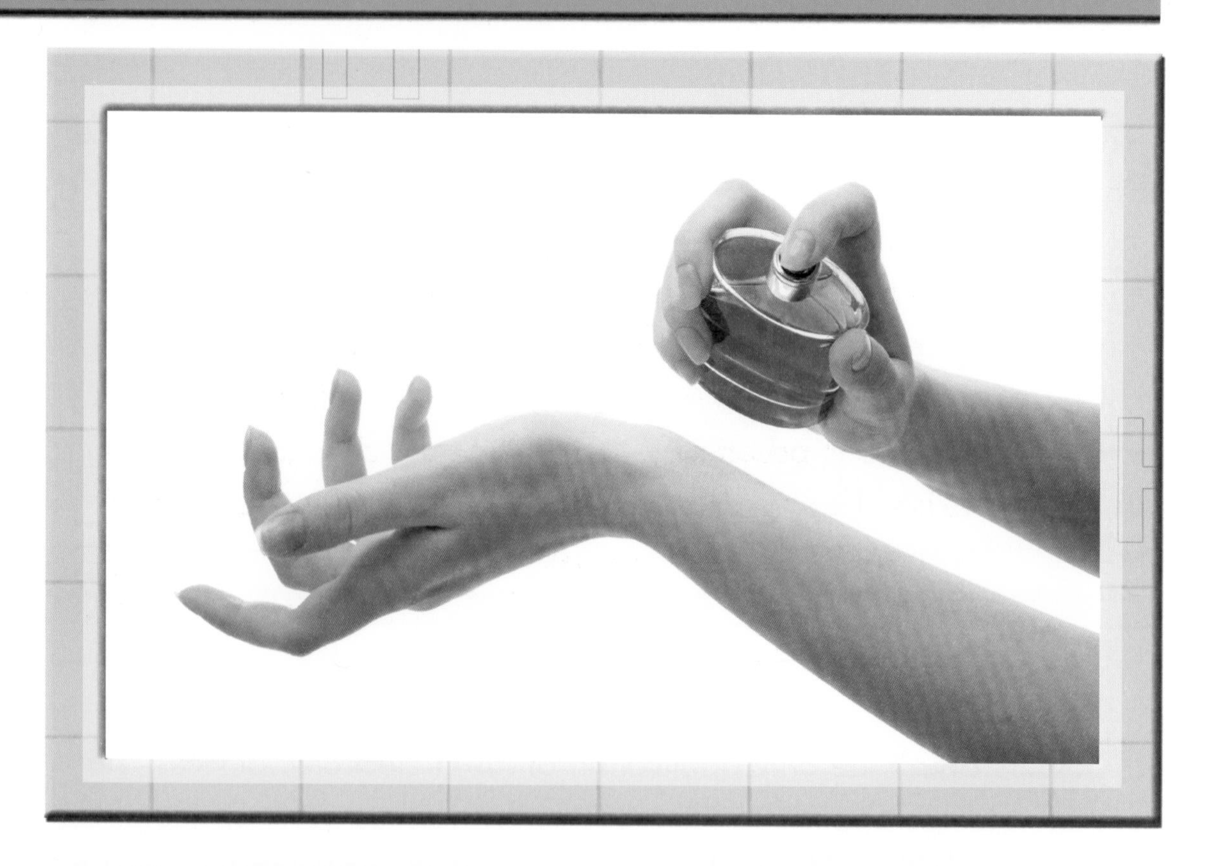

Perfumes basically have three parts. One part is the perfume oils, which provide the fragrance. Another part is something like alcohol, into which the perfume oils are dissolved. Then there are stabilizers such as tin(II) chloride and tin(IV) chloride.

steel coated with tin. Did you use safety pins or straight pins at home? They're probably coated with tin, too.

Inorganic tin compounds also play an important role at home and in school. If you read the labels of canned and bottled foods and fruit juices, you may discover tin(II) chloride among the ingredients. Tin(II) chloride is often added to prevent reactions of the food with oxygen and help the food or juice keep its color. It's used as a stabilizer in soaps and perfumes as well. So is tin(IV) chloride. Think of these two tin chloride compounds the next time you wash your hands or spray yourself with your favorite scent.

Another valuable inorganic tin compound is tin(II) fluoride (SnF_2). In small amounts, fluoride ions (F^-) help strengthen teeth to prevent tooth decay. Therefore, tin(II) fluoride is one of several fluoride compounds you'll find in toothpastes and dental rinses.

As you've learned, tin has played a prominent role in human society for 5,500 years and continues to be important in modern life. Our lives would be very different without this adaptable metal. Yet, you've also learned that tin ore is becoming scarce. We can't change the amount of tin ore that exists. However, we can help ensure that we have tin for all the metal's varied applications by recycling products containing the element. That way, future generations will be able to continue to enjoy the benefits of this amazing metal.

The Periodic Table of Elements

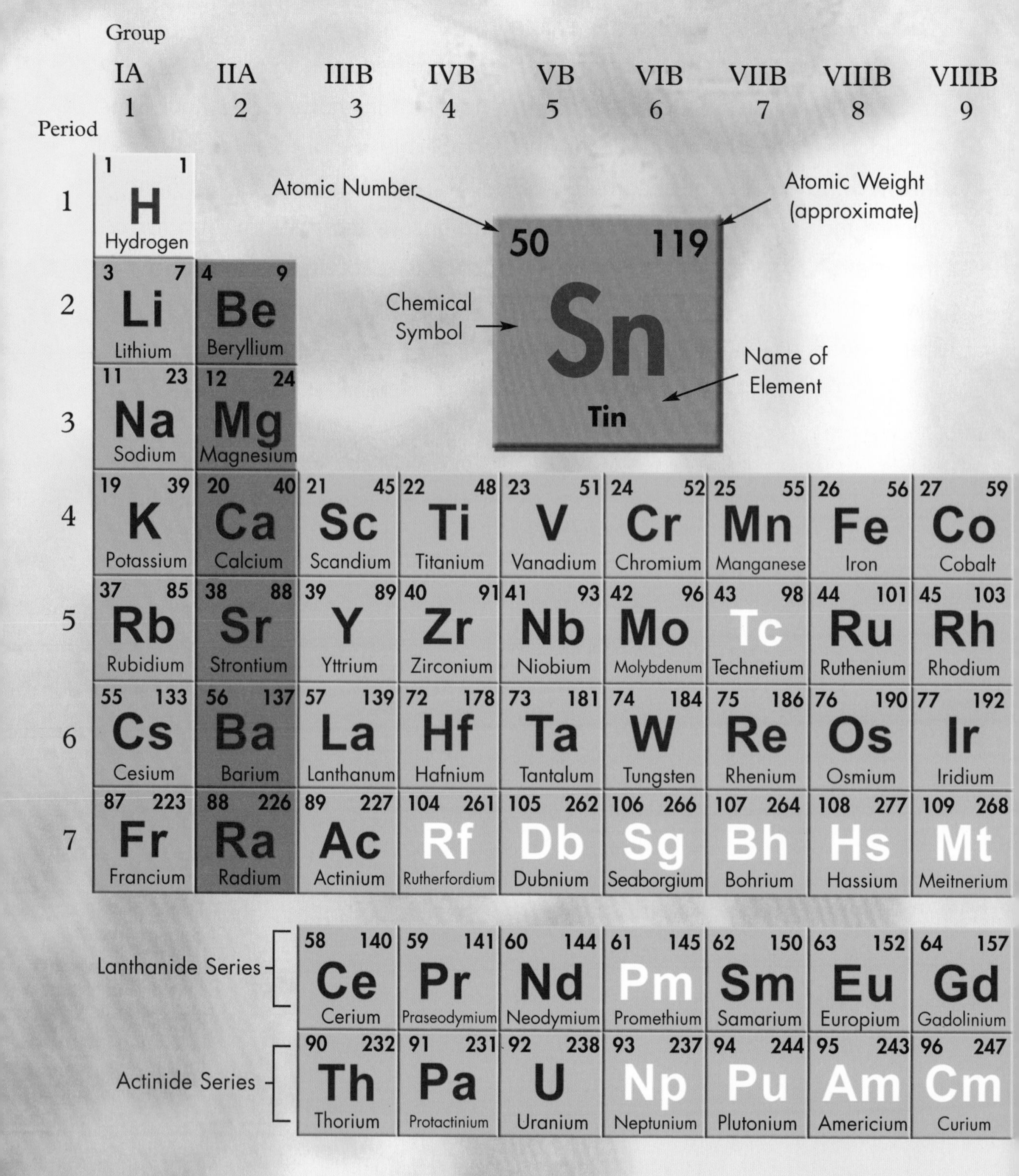

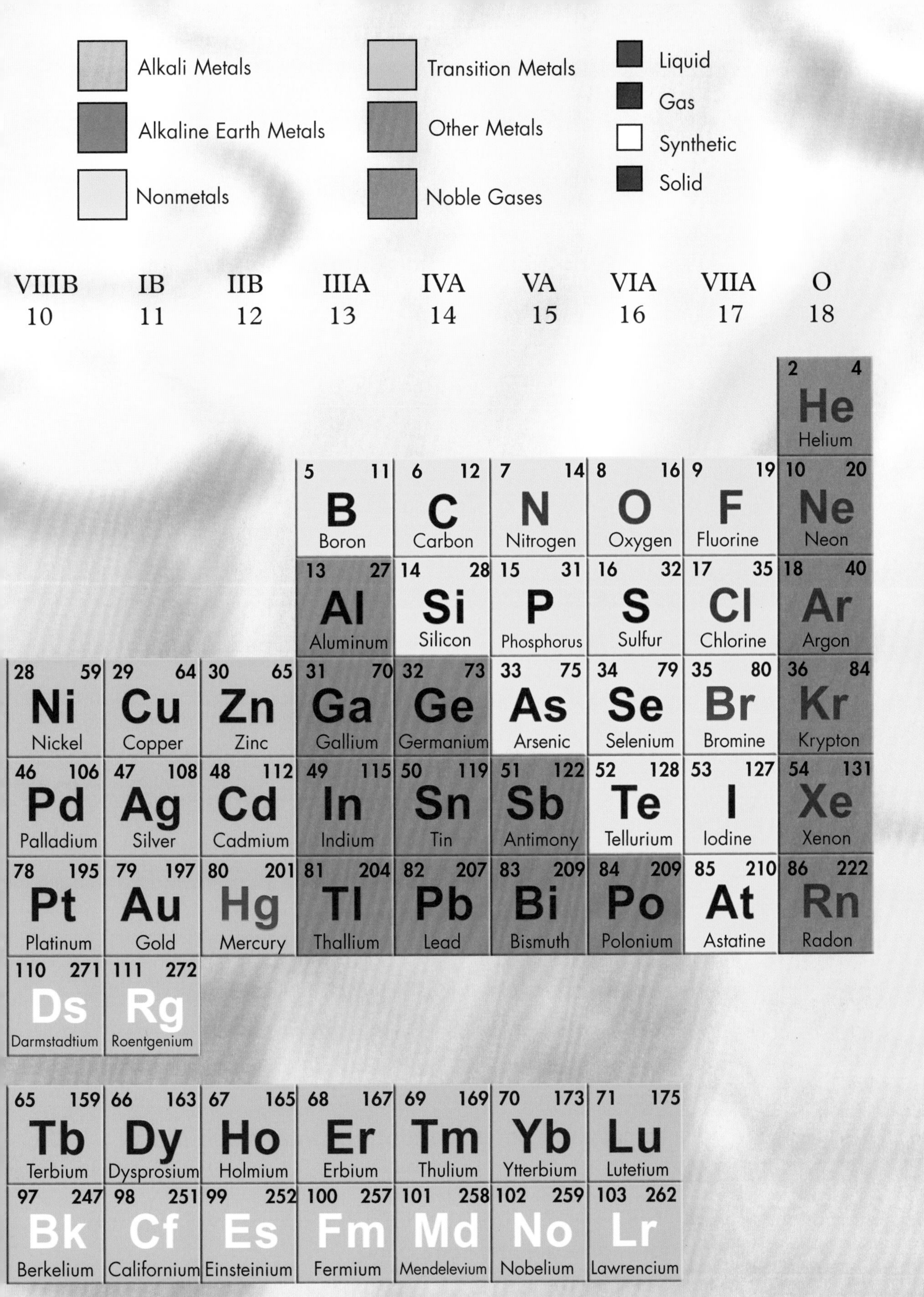
Alkali Metals
Alkaline Earth Metals
Nonmetals
Transition Metals
Other Metals
Noble Gases
Liquid
Gas
Synthetic
Solid
VIIIB 10
IB 11
IIB 12
IIIA 13
IVA 14
VA 15
VIA 16
VIIA 17
O 18
2 4 He Helium
5 11 B Boron
6 12 C Carbon
7 14 N Nitrogen
8 16 O Oxygen
9 19 F Fluorine
10 20 Ne Neon
13 27 Al Aluminum
14 28 Si Silicon
15 31 P Phosphorus
16 32 S Sulfur
17 35 Cl Chlorine
18 40 Ar Argon
28 59 Ni Nickel
29 64 Cu Copper
30 65 Zn Zinc
31 70 Ga Gallium
32 73 Ge Germanium
33 75 As Arsenic
34 79 Se Selenium
35 80 Br Bromine
36 84 Kr Krypton
46 106 Pd Palladium
47 108 Ag Silver
48 112 Cd Cadmium
49 115 In Indium
50 119 Sn Tin
51 122 Sb Antimony
52 128 Te Tellurium
53 127 I Iodine
54 131 Xe Xenon
78 195 Pt Platinum
79 197 Au Gold
80 201 Hg Mercury
81 204 Tl Thallium
82 207 Pb Lead
83 209 Bi Bismuth
84 209 Po Polonium
85 210 At Astatine
86 222 Rn Radon
110 271 Ds Darmstadtium
111 272 Rg Roentgenium
65 159 Tb Terbium
66 163 Dy Dysprosium
67 165 Ho Holmium
68 167 Er Erbium
69 169 Tm Thulium
70 173 Yb Ytterbium
71 175 Lu Lutetium
97 247 Bk Berkelium
98 251 Cf Californium
99 252 Es Einsteinium
100 257 Fm Fermium
101 258 Md Mendelevium
102 259 No Nobelium
103 262 Lr Lawrencium

Glossary

allotrope One of two or more forms of the same state (solid, liquid, or gas) that some elements have that differ in the bonding arrangements of the atoms.

alloy A solid mixture composed of two or more elements with at least one metal.

atom The smallest part of an element having the chemical properties of that element.

biocide A substance that kills many kinds of living things.

corrosion The act or process of wearing something away little by little by the chemical reaction of that substance with species in its surroundings, such as iron metal reacting with oxygen to form rust.

crystalline Formed in a regular pattern.

fire retardant A substance that helps prevent materials from catching fire and helps slow the spread of fire.

flux A substance added to ore during smelting to improve the flow of the melt and/or to remove impurities.

immune system The system in the body that helps protect it from things that can cause disease.

impurity An unwanted substance in another substance.

ion An atom or molecule that has unequal numbers of protons and electrons.

isotopes Atoms containing the same number of protons but different numbers of neutrons.

leach To remove something from a substance by means of a liquid passed through it.

mordant A chemical that fixes a dye in or on a substance by combining with the dye to form an insoluble compound.

oxidation state The term used to describe how many electrons an atom has lost or gained from its original state.

oxidize To combine with oxygen; to lose electrons.

plate To cover something with a layer that sticks to it.

property One of the chemical or physical characteristics of an element or compound.

radioactive Having the property of giving off subatomic particles and/or energy as the result of a reaction of atomic nuclei.

recycle To reprocess something old to use it again in a new item.

reverberatory furnace A furnace that uses hot combustion gases to melt the ore and flux materials.

sensor A device that detects something such as a gas, heat, light, or sound and reacts to it.

smelt To chemically react an ore in order to produce the metal, usually at high temperatures in a melted ore mixture.

stabilizer A substance added to another substance to prevent or slow an unwanted change.

superconductive Having no resistance to electricity at all, so that all electricity flowing through it is transmitted and none is lost to unwanted heat production.

For More Information

Agency for Toxic Substances and Disease Registry (ATSDR)
Department of Health and Human Services
1825 Century Boulevard
Atlanta, GA 30345
(800) 232-4636
Web site: http://www.atsdr.cdc.gov
The ATSDR is a federal public health agency that works to prevent harmful exposures and diseases related to toxic substances by using science, taking public health actions, and providing health information. Tin compounds are among the substances the agency is concerned with.

American Chemical Society
1155 Sixteenth Street NW
Washington, DC 20036
(800) 227-5558
Web site: http://www.chemistry.org/portal/a/c/s/1/home.html
The American Chemical Society is the national organization for professional chemists. It also provides information about all aspects of chemistry for students and educators.

ITRI, Ltd.
Unit 3, Curo Park
Frogmore, St. Albans
Hertfordshire AL2 2DD
United Kingdom
Web site: http://www.itri.co.uk

ITRI is an organization of tin-producing companies that was formed in 1932. Dedicated to supporting the tin industry and expanding tin use, it is a source for a wide range of information about tin.

The Minerals, Metals and Materials Society (TMS)
184 Thorn Hill Road
Warrendale, PA 15086
(724) 776-9000
Web site: http://www.tms.org
TMS is an international professional organization for individuals involved in minerals processing, primary metals production, basic research, and the advanced applications of materials.

Tin Stabilizers Association (TSA)
100 North 20th Street, 4th Floor
Philadelphia, PA 19103
(215) 564-3484
Web site: http://www.tinstabilizers.com
The TSA is a nonprofit organization that promotes the use of tin stabilizers in vinyl applications. It provides information about tin stabilizers and cooperates with other organizations on scientific studies.

Web Sites

Due to the changing nature of Internet links, Rosen Publishing has developed an online list of Web sites related to the subject of this book. This site is updated regularly. Please use this link to access the list:

http://www.rosenlinks.com/uept/tin

For Further Reading

Diagram Group, The. *The Facts On File Chemistry Handbook*. Rev. ed. New York, NY: Facts On File, 2006.

Emsley, John. *Nature's Building Blocks: An A–Z Guide to the Elements*. New York, NY: Oxford University Press, 2002.

Ganeri, Anita. *Carbon and the Group 4 Elements* (The Periodic Table). Chicago, IL: Heinemann Library, 2004.

Gray, Leon. *Tin* (The Elements). Tarrytown, NY: Benchmark Books, 2004.

Keller, Rebecca W. *Chemistry: Level 1* (Real Science-4-Kids). Albuquerque, NM: Gravitas Publications, Inc., 2005.

Miller, Ron. *The Elements: What You Really Want to Know*. Minneapolis, MN: 21st Century, 2006.

Oxlade, Chris. *Elements and Compounds* (Chemicals in Action). Rev. updated ed. Chicago, IL: Heinemann Library, 2007.

Saunders, Nigel. *Carbon and the Group 14 Elements* (The Periodic Table). Chicago, IL: Heinemann Library, 2003.

Stwertka, Albert. *A Guide to the Elements*. 2nd ed. New York, NY: Oxford University Press, 2002.

Tocci, Salvatore. *Tin* (A True Book). Danbury, CT: Children's Press, 2005.

Wertheim, Jane. *The Usborne Illustrated Dictionary of Chemistry*. Rev. ed. London, England: Usborne Books, 2008.

Bibliography

Agency for Toxic Substances and Disease Registry (ATSDR). "ToxFAQs for Tin." September 11, 2007. Retrieved March 21, 2008 (http://www.atsdr.cdc.gov/tfacts55.html).

Carlin Jr., James F. *Tin Recycling in the United States in 1998*. U.S. Department of the Interior, U.S. Geological Survey. Retrieved March 23, 2008 (http://pubs.usgs.gov/of/2001/of01-433/of01-433.pdf).

Ede, Andrew. *The Chemical Element: A Historical Perspective*. Westport, CT: Greenwood Press, 2006.

Emsley, John. *Nature's Building Blocks: An A–Z Guide to the Elements*. New York, NY: Oxford University Press, 2002.

Greenwood, N. N., and A. Earnshaw. *Chemistry of the Elements*. 2nd ed. Woburn, MA: Butterworth-Heinemann, 1997.

ITRI, Ltd. "Tin Use Survey 2007." 2008. Retrieved March 23, 2008 (http://www.itri.co.uk/pooled/articles/BF_TECHART/view.asp?Q=BF_TECHART_297350).

Key to Metals Task Force. "Tin and Tin Alloys." Key to Metals. Retrieved March 24, 2008 (http://www.key-to-metals.com/Article26.htm).

Krebs, Robert E. *The History and Use of Our Earth's Chemical Elements*. 2nd ed. Westport, CT: Greenwood Press, 2006.

Lenntech. "Tin (Sn)." Retrieved March 21, 2008 (http://www.lenntech.com/Periodic-chart-elements/Sn-en.htm).

Martin, Holly Bigelow. "An Introduction to Tin Whiskers: From Pacemakers to Power Plants, Tiny Metal Growths Cause Problems." Suite101.com, November 29, 2007. Retrieved March 30, 2008 (http://engineering.suite101.com/article.cfm/an_introduction_to_tin_whiskers).

New Scientist. "How Long Will It Last?" Vol. 194, No. 2605. May 26, 2007, pp. 39–39.

Rovner, Sophie L. "Stopping the Tin Whisker Stalkers." *Chemical & Engineering News*, Vol. 85, No. 29. July 16, 2007, pp. 30–33.

Stwertka, Albert. *A Guide to the Elements.* 2nd ed. New York, NY: Oxford University Press, 2002.

Thompson, David W. "Tin." *World Book Multimedia Encyclopedia.* Chicago, IL: World Book, Inc., 2001.

Index

About the Author

Janey Levy is a writer and editor who has written more than seventy-five books for young people. Her interest in chemistry comes from her curiosity about the role the body's own chemistry plays in an individual's physical, emotional, and mental well-being. She has written about other elements of the periodic table, including krypton and radon. Levy lives in Colden, New York.

Photo Credits

Cover, pp. 1, 13, 17, 38–39 by Tahara Anderson; pp. 5, 29 © Charles D. Winters/Photo Researchers, Inc.; p. 6 © The British Museum/Topham/The Image Works; p. 8 © Science Museum/SSPL/The Image Works; p. 10 © Royal Society of Chemistry; p. 18 © sciencephotos/Alamy; p. 20 © Renee Purse/Photo Researchers, Inc.; p. 22 © Guy Newman/Apex News and Pictures Agency/Alamy; p. 23 © SSPL/The Image Works; p. 25 NASA/Wikipedia; p. 26 © www.istockphoto.com/robert lerich; p. 28 B2M Productions/Digital Vision/Getty Images; p. 31 Chief Photographer's Mate Dennis Taylor/U.S. Navy; p. 33 © www.istockphoto.com/Arno Massee; p. 34 © www.istockphoto.com/angelhell; p. 36 © www.istockphoto.com/Sergey Kishan.

Designer: Tahara Anderson; **Editor:** Kathy Kuhtz Campbell
Photo Researcher: Amy Feinberg